Math Mammoth
Grade 4
Skills Review Workbook
Answer Key

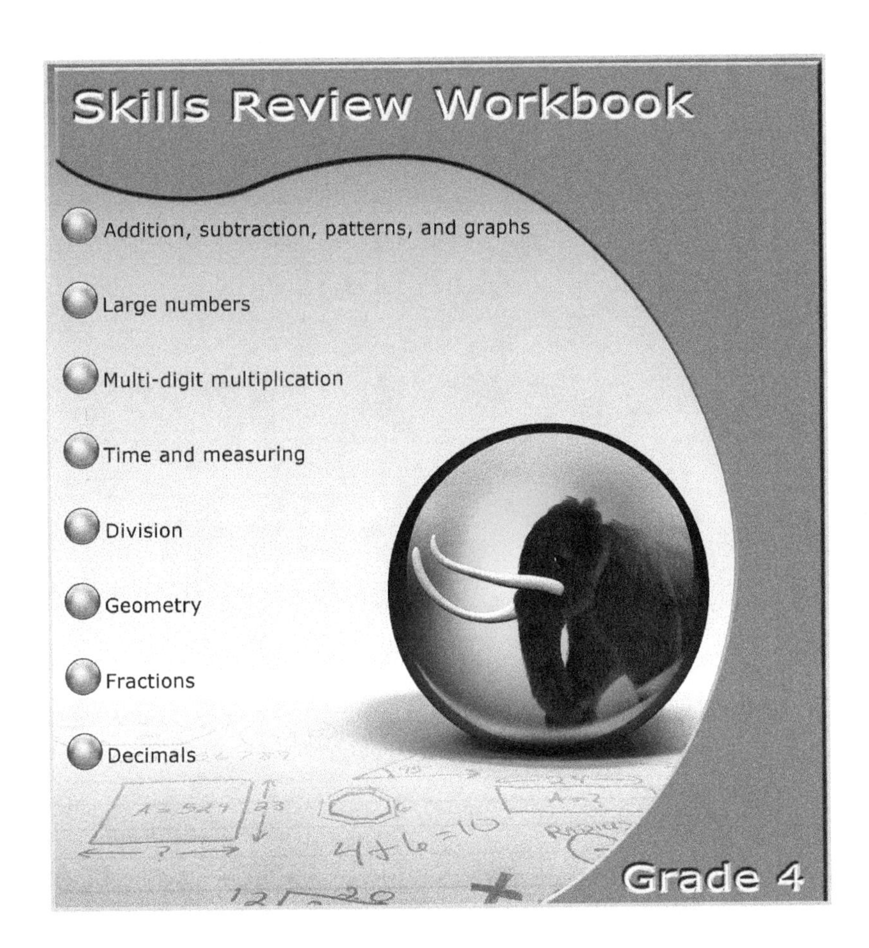

By Maria Miller

ISBN 978-1-942715-41-2

Edition 5/2018

Contents

Chapter 1: Addition, Subtraction, Patterns and Graphs

Skills Review 1, p. 7

1. a. 160 b. 81 c. 123

2.

1,250	1,410	1,840	1,120	1,680
640	710	930	840	860
310	230	460	540	420
120	160	240	190	210
64	85	122	105	100

3. a. 296 b. 694 c. 891

4. The total cost of the car was $2,715 + $1,847 + $2,069 = $6,631.

5. a. She gave Mrs. Harrison 24 cookies. The easiest way to solve this mentally might be to think how much to add to 48 so that you get 72.
   b. Nineteen cookies had been eaten. The easiest way to solve this mentally might be to think how much to add to 29 so that you get 48.

Puzzle corner

$$\begin{array}{r} 5636 \\ +\ 2966 \\ \hline 8602 \end{array} \qquad \begin{array}{r} 2676 \\ +\ 1378 \\ \hline 4054 \end{array}$$

Skills Review 2, p. 8

1. a. 236; 236 + 469 = 705
   b. 3,415; 3415 + 5835 = 9250
   c. 5272; 5272 + 741 = 6013

2. $190 + $45 = $235 Yes, she has enough money.
   $250 − $235 = $5. She has $5 left.

3. a. 209 b. 620

4. a. $1150 = 1150$ b. $6800 > 6600$
   c. $1500 > 1400$ d. $1320 < 1420$

5. a. 16, 81, 60
   b. 6, 110, 15
   c. 56, 12, 48
   d. 0, 60, 55

Skills Review 3, p. 9

1. He traveled 1,958 + 2,091 + 4,600 = 8,649 total miles.

2.

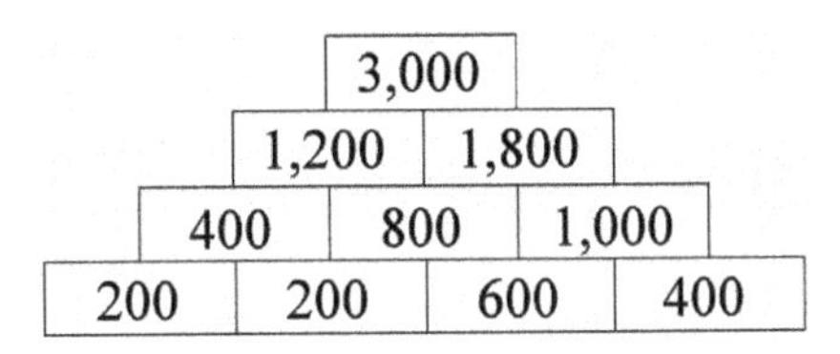

3. a. 4 b. 5 c. 6 d. 6

4.

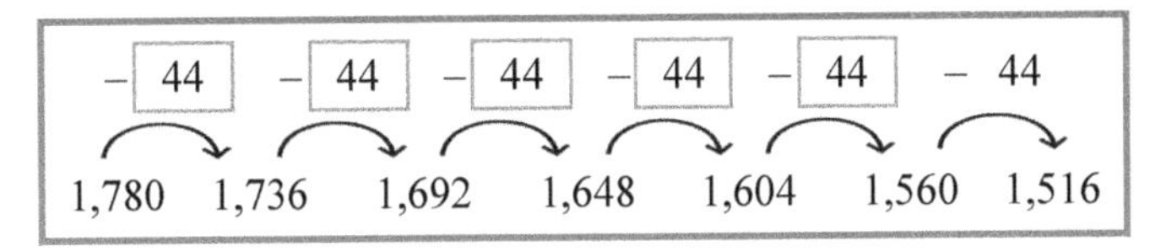

5. The third flock has 1,348 − 430 − 508 = 410 sheep.

Skills Review 4, p. 10

1. a. 1,585; 1,585 + 4,820 + 695 = 7,100
   b. 2,551; 2,551 + 463 + 78 = 3,092

2.

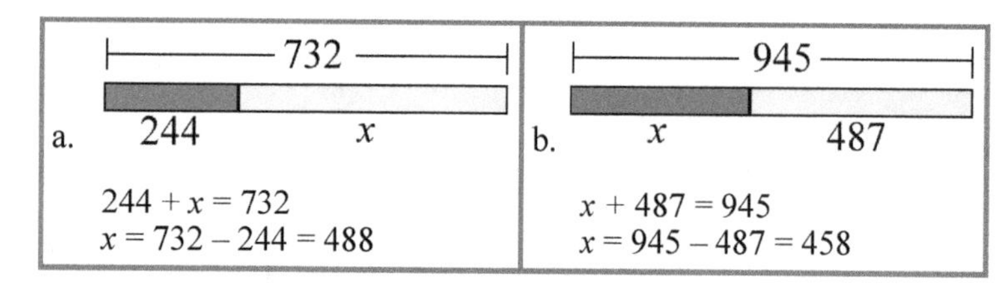

3. 

4. a. 7,000 + 900 + 40 + 8
   b. 3,000 + 90 + 2

Skills Review 5, p. 11

1. a. 48 + 4 = 52   b. 3 × 8 ÷ 4 = 6   c. 90 – 49 = 41

2. a. 1,310   b. 1,594   c. 657

3. *Instead of adding 69, add **70**, and then subtract **1**.*

n	730	690	650	610	570	530	490	450
$n + 69$	799	759	719	679	639	599	559	519

4. a. 54 ÷ 6 = 9; Each child got nine cherries.
   b. 54 – 6 × 4 = 30 OR 6 × 4 = 24 and 54 – 24 = 30; They saved a total of 30 cherries to eat later.

5. a. 10   b. 9   c. 12

Puzzle corner: 3 × 9 + 7 × 8 – (3 + 7) = 73 OR 3 × 9 + 7 × 8 – 3 – 7 = 73. There were 73 chairs left.

Skills Review 6, p. 12

1. 208; 208 + 495 = 703

2.

×	4	7	9	6	8	12
6	24	42	54	36	48	72
8	32	56	72	48	64	96
4	16	28	36	24	32	48
9	36	63	81	54	72	108

3.

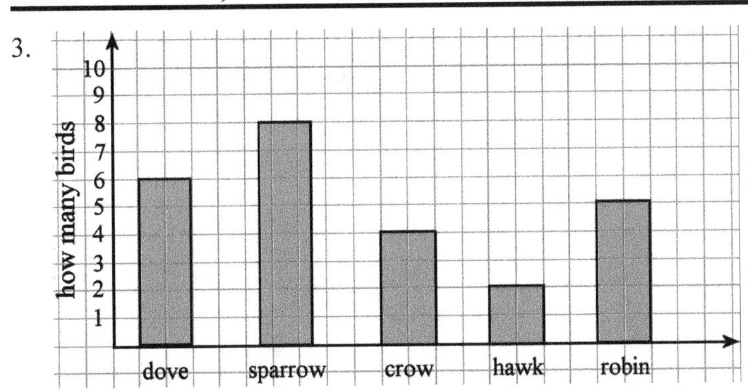

Bird	Frequency
dove	6
sparrow	8
crow	4
hawk	2
robin	5

Puzzle corner: There are many possible solutions. One possibility is below:

3,780	–	20	+	50	= 3,810
–		+		–	
50	+	40	+	30	= 120
+		–		+	
20	+	50	+	130	= 200
=		=		=	
3,750		10		150	

Skills Review 7, p. 13

1. a. $x - 95 = 238;\ \ x = \$333$
   b. $128 + 33 = x;\ \ x = 161$

2. 4,215 – 1,346 – 2 = 2,867. There are <u>2,867 passengers</u>.

3. a. Becky earned the most money during the month of August.
   b. She earned a total of \$120 + \$140 + \$160 = <u>\$420</u> during those three months.

4. The numbers highlighted in blue have been corrected.

n	330	370	410	450	490	530	570	610
$n + 79$	409	449	489	529	569	609	649	689

5.

```
  1 1 2
  6 8 3 0
  1 5 9 7
    3 0 5
+     2 8
---------
  8 7 6 0
```

Skills Review 8, p. 14

1.

n	95	4,762	344
nearest 10	100	4,760	340
nearest 100	100	4,800	300
nearest 1000	0	5,000	0

2.

14 – 6 = 8
64 – 6 = 58
142 – 60 = 82
1,430 – 600 = 830

3. Answers will vary. Please check the student's work. The following is just one possible answer.

22 – 7 = 15   62 – 7 = 55
132 – 7 = 125   220 – 70 = 150
820 – 70 = 750   4,200 – 700 = 3,500

4. a. $9 \times 7 + 8 = 71$ b. $8 \times (32 - 29) - 24 = 0$
   c. $(7 \times 8) \div 4 = 14$ d. $6 \times 12 + (41 - 9) = 104$

5. a. $42 \div 6 = 7$ b. $100 \div 10 = 10$
   c. $36 \div 4 = 9$ d. $56 \div 8 = 7$
   e. $15 \div 3 = 5$

6. 2,590; 2,590 + 3,975 + 648 = 7,213

Skills Review 9, p. 15

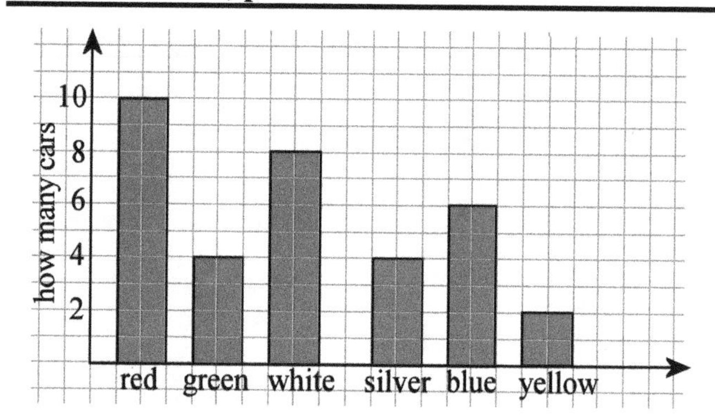

Color	Frequency
Red	10
Green	4
White	8
Silver	4
Blue	6
Yellow	2

2. Estimate: 4,100 – 1,000 – 300 = 2,800 Exact: 2,786

3. a. She had to drive a total of 320 + 290 + 865 = 1,475 miles.
   b. She drove 1,475 + 1,475 = 2,950 miles round-trip.

Skills Review 10, p. 16

1. a. Discount \$8.50 b. New price \$19.48 c. Original \$21.32

2. a. $2 \times 8 + 14 = 30$. Their total weight is 30 pounds.
   b. $5 \times 10 + 2 \times 5 = 60$. He practices 60 minutes each week.

3. Please check the word problem the student created. Answers will vary. $x = 2{,}920$

4.

n	360	440	520	600	680	760	840	920
$n - 59$	301	381	461	541	621	701	781	861

Skills Review 11, p. 17

1. a. 198; 3,504 + 198 = 3,702 b. 1,085; 976 + 1,085 = 2,061 c. 357; 183 + 357 = 540

2. a. 435 b. 40 c. 60

3.

Number	70	230	350	800	2,700	5,500
Double the number	140	460	700	1,600	5,400	11,000

4. a. 1,000 b. 0 c. 4,000

Puzzle corner:

7	×	6	= 42
×		×	
9	×	4	= 36
=		=	
63		24	

OR

21	×	2	= 42
×		×	
3	×	12	= 36
=		=	
63		24	

3	×	7	= 21
×		×	
5	×	8	= 40
=		=	
15		56	

Chapter 2: Large Numbers and Place Value

Skills Review 12, p. 18

1. Please check the students work. See the example below.

400	200	700	150	500
300	600	900	400	550
900	750	300	600	200
400	100	450	500	800
250	300	600	750	300

2. a. 428 b. 384 c. 480 d. 855

3. Answers may vary according to how the numbers are rounded.
   a. $48 – $24 = $24. Chad needs about $24 more. b. $16 + $9 = $25. They cost about $25.

4. a. 12 b. 9 c. 8

5. Check the student's answer. Answers can vary.
   a. 64 ÷ 7 = 9 R1. Jim can weed 9 ft on each of the six days, and 10 ft on the seventh day.
   b. 64 ÷ 6 = 10 R4. He can weed 11 ft on each of the five days and 9 ft on the sixth day.

6. a. $4 \times 6 = 24$ b. $49 + 3 = 52$ c. $12 \times 6 + 7 = 79$

Skills Review 13, p. 19

1. a. 600 b. 8,000 c. 3 d. 900

2. a. 10,202 b. 12,415

3. a.

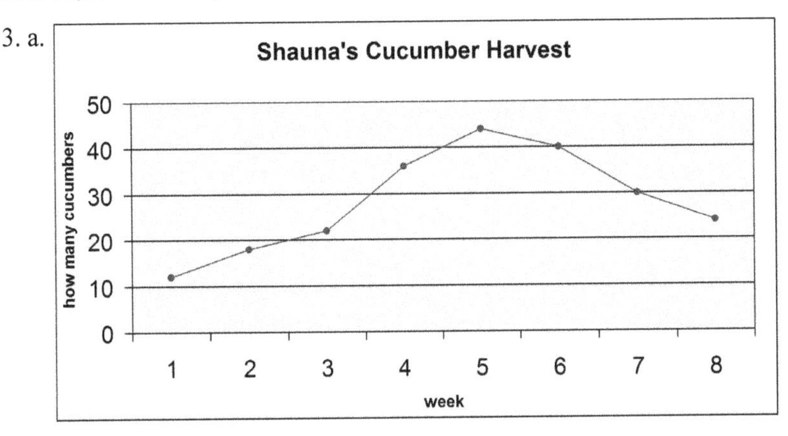

   b. She harvested a total of 226 cucumbers.

4. $314 - x = 190$ OR $314 - 190 = x$; $x = 124$. There were 124 people who did not attend.

Skills Review 14, p. 20

1. a. 6,380 + 620 = 7,000 b. 9,730 + 270 = 10,000 c. 7,520 + 480 = 8,000

2. She can buy three packages, so she can get $3 \times 6 = 18$ apples.

3. a. 12 b. 6 c. 6 d. 3

4. 530 + 630 + 420 + 580 ≈ 2,160 miles.

Skills Review 14, cont.

5. Answers will vary. Please check the student's work. See an example below.

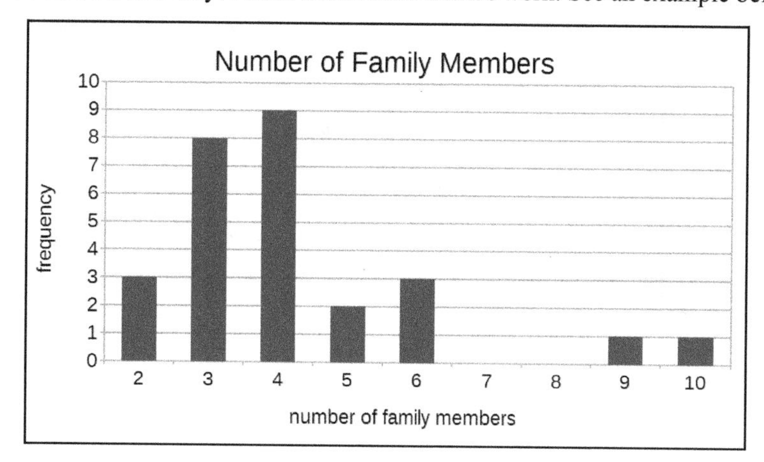

Size of family	Frequency
2	3
3	8
4	9
5	2
6	3
7	0
8	0
9	1
10	1

c. The most common size of family was four. Nine friends had that size of family.

Skills Review 15, p. 21

1. a. 17 b. 58 c. 20

2. a. 19,000 b. 533,000 c. 720,000 d. 891,000

3.

Item Cost	Given	Change	$5 bill	$1 bill	25¢	10¢	5¢	1¢
a. $11.56	$20	$8.44	1	3	1	1	1	4
b. $8.33	$10	$1.67	0	1	2	1	1	2
c. $3.44	$5	$1.56		1	2		1	1
d. $13.97	$20	$6.03	1	1				3

4. a. 3 b. 12 c. 8 d. 6 e. 5 f. 8

5. $2 \times 243 + 58 = 544$ stamps in total.

6. a. 37 b. 85

Skills Review 16, p. 22

1. a. $70,000 + 9,000 + 300 + 10 + 2$
   b. $600,000 + 5,000 + 400 + 80 + 3$
   c. $10,000 + 7,000 + 600 + 4$

2. a. \$55.38; \$55.38 + \$38.62 = \$94.00 b. \$180.53; \$180.53 + \$619.47 = \$800.00

3. Answers will vary. Please check the student's work. See examples below:
   What flavor of pie was the least popular? How many people took the survey?
   How many people liked apple pie better than custard? What flavor did they like the best?

4.

n	**72**	**7,389**	**12,706**
nearest 10	70	7,390	12,710
nearest 100	100	7,400	12,700
nearest 1000	0	7,000	13,000

5. a. 145 b. 144 c. 61

Skills Review 17, p. 23

1.

183

x	50	70

Addition:
$x + 50 + 70 + = 183$
Solution: $x = 63$

2.

a. 5,700	b. 1,260
6,100	1,180
6,500	1,100
6,900	1,020
7,300	940
7,700	860
8,100	780

3. a. 98,600 > 86,900	b. 13,040 < 13,400	c. 224,920 < 242,290
d. 59,060 > 59,006	e. 604,312 < 640,123	f. 450,083 > 45,830

4. a. Clarissa spends about 4 × \$160 = \$640. Ashley spends about 4 × \$170 = \$680.
   b. The difference is about \$40.

5. a. 7,975 b. 10,994 c. 1,970

Skills Review 18, p. 24

1. a. 782,108 b. 151,909 c. 721,229

2.

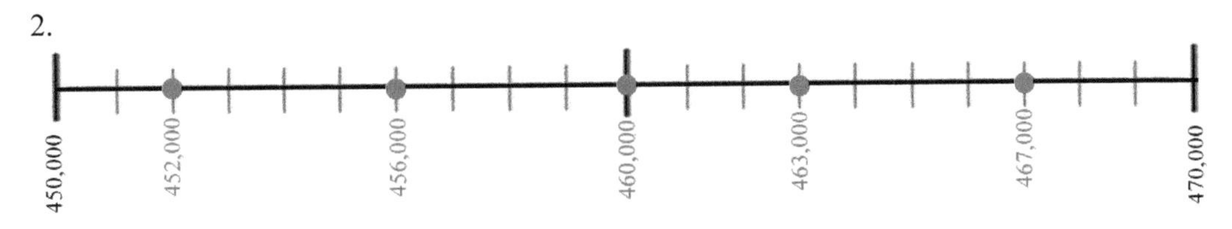

3. a. $18 + 9 \times 2 = 36$ b. $74 - (12 \times 6) - 2 = 0$ c. $8 \times 7 - 9 = 47$

4. a. about 140 people
   b. March, April and November
   c. June, July, August, September
   d. Answers can vary slightly. There were about 1,900 visitors during the year.

Skills Review 19, p. 25

1.

number	195,212	422,783
to the nearest 1,000	195,000	423,000
to the nearest 10,000	200,000	420,000
to the nearest 100,000	200,000	400,000

2.

a. $\underline{870} - 90 = 780$
b. $460 + 370 = \underline{830}$
c. $920 - \underline{40} = 880$
d. $549 + \underline{70} = 619$

3.

a. $6 \times 6 = 36$	b. $4 \times 11 = 44$	c. $9 \times 8 = 72$
d. $12 \times 12 = 144$	e. $3 \times 5 = 15$	f. $0 \times 10 = 0$

4.

a. \$18.50 – \$8.65 = x OR \$8.65 + x = \$18.50
   x = \$9.85 The skirt cost \$9.85.

b. x – \$7.35 – 3 × \$2.68 = \$4.73
   (Or possibly \$7.35 + 3 × \$2.68 + \$4.73 = x)
   x = \$20.12 Originally, he had \$20.12.

5.

Estimate:	Exact:
4,182 + 539 – 706 ↓ ↓ ≈ 4,200 + 500 – 700 = 4,000	4,015

Skills Review 20, p. 26

1.

+ 230	+ 230	+ 230	+ 230	+ 230	+ 230	
2360	2590	2820	3050	3280	3510	3740

2. $320 - 4 \times 40 = 160$. She has 160 pages left to read.

3. $150 – 4 × $8 – 3 × $7 = $97. He has $97 left.

4. a. 265; $265 + 597 = 862$ b. 2,788; $2788 + 4246 = 7034$ c. 415; $415 + 805 = 1220$

5. a. 7,531   b. 1,357   c. 6,174

6. a. 11   b. 4   c. 4   d. 9   e. 4   f. 5

Chapter 3: Multi-Digit Multiplication

Skills Review 21, p. 27

1.

5,380			
3,860	1,520		
2,910	950	570	
2,240	670	280	290

2. a. 2,960 b. 6,992 c. 1,930 d. 8,975

3. a.

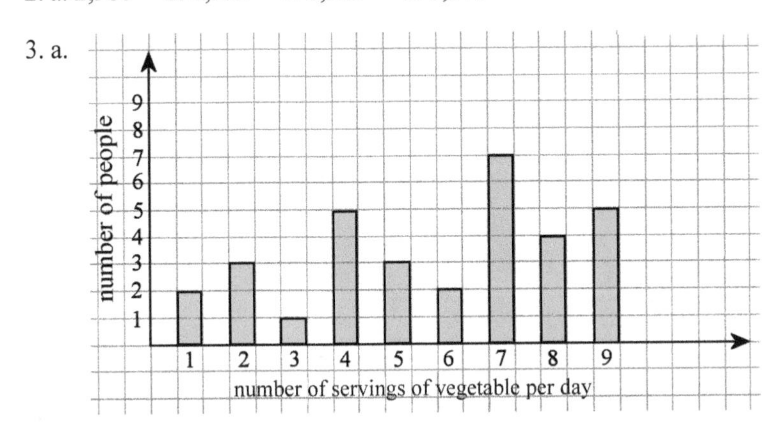

Veggies per day	Number of people
1	2
2	3
3	1
4	5
5	3
6	2
7	7
8	4
9	5

b. Fourteen people ate five or less vegetables per day.

4. a. 5,321 + 5,321 = 10,642 < 14,000. Yes, the number of people more than doubled.
b. 14,000 − 10,642 = 3,358. There were 3,358 more people than double the original population.

Skills Review 22, p. 28

1. a. 865,866 b. 214,373 c. 924,755

2. a. 71,400 b. 27,300 c. 899,000

3. $x = 647$

4. 430 < 3,040 < 3,400 < 4,003 < 4,030 < 4,300

5. Answers may vary. For example: 3 × \$45 ≈ \$135 or 3 × \$40 ≈ \$120 or 3 × \$44 ≈ \$132.

6. a. 53 b. 1,200 c. 674

7. *Hint: At first add 2,000, instead of 1,999.*

n	238	572	891	1,260	1,647	2,033
n + 1,999	2,237	2,571	2,890	3,259	3,646	4,032

Skills Review 23, p. 29

1. a. $2 \times 7 + 2 \times 8 = 30$ or $2 \times (7 + 8) = 30$. They picked a total of 30 flowers.
   b. $9 \times 6 - (4 \times 3 + 3) = 39$. She had 39 cookies left.

2. a. $3,625 b. $92 c. $706

3. a. $x - 280 = 824$; $x = 1,104$ b. $1,048 - 490 = x$ OR $1,048 - x = 490$; $x = 558$

4. a. $4670 - 4425 = 245$ The difference was 245 points.
   b. $4900 - 4670 = 230$; Marcella needed 230 more points plus one to win the game.

Skills Review 24, p. 30

1. a. 6, 6 b. 9, 5 c. 12, 4 d. 7, 8

2. $4,924 – $1195 – $1428 = $2,301; Robert paid $2,301.

3. a. $4670 + 330 = 5,000$ b. $2420 + 580 = 3,000$ c. $9780 + 220 = 10,000$

4. a. $62 \div 9 = 6$ R8. Madeline got six bags of tomatoes.
   b. She used eight tomatoes to make salsa.

5. a. 219 b. 1,013

6.

Number	Rounded number	Rounding error
7,489	7,000	489
2,940	3,000	60
5,720	6,000	280

7.

a. $600 - 8 = 592$
b. $340 - 70 = 270$
c. $53 - 17 = 36$
d. $2,000 - 19 = 1981$

Skills Review 25, p. 31

1. $7460 – x = $6945; x = $515. The discount was $515.

← original price $7,460 →

$6,945	$515

2. a. 608,043 b. 93,058

3. a. 22 b. 40 c. 6

4. 415,374

5. a. $76 \div 8 = 9$ R4. There were nine full tables.
   b. There were four people.

6.

Person	**Jessica**	**Dawson**	**Vanessa**
Amount of money to spend	$326	$413	$204
How much money spent	$78	$95	$46
How much money left	$248	$318	$158

Skills Review 26, p. 32

1. a. 60 b. 11 c. 90

2. a.

how many eggs: 20, 18, 16, 14, 12, 10, 8, 6, 4, 2

Mon Tue Wed Thu Fri Sat Sun

days

b. She gathered 82 eggs.

3. a. That means about \$36,000. So, he pays about \$3,000 monthly.
   b. That is about 120,000 words in the whole book.

4. a. Check the student's answer. \$28 + 3 × \$4.35 = \$41.05.
   b. Check the student's answer. \$36 −3 × \$2.68 = \$27.96.

Skills Review 27, p. 33

1.

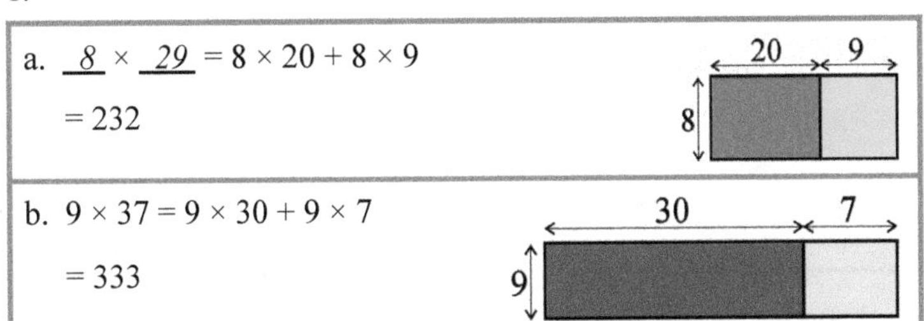

2. a. Robert flew 2 × 12,181 = 24,362 miles. Candace flew 5,494 + 19,081 = 24,575 miles. <u>Candace</u> traveled more miles.
   b. She flew 213 more miles.

3. a. < b. > c. < d. < e. < f. <

4.

+600 +600 +600 +600 +600 +600

5,600 6,200 6,800 7,400 8,000 8,600 9,200

Skills Review 28, p. 34

1. You want to round one factor up and the other down, if possible, and definitely not round both factors down, as you would then be underestimating your cost. Answers will vary according to how the student chooses to round. Please check the student's answers. Some possible examples:
   a. $20 \times \$7.00 = \$140$
   b. $13 \times \$300 = \$3,900$ OR $12 \times \$350 = \$3,500 + \$700 = \$4,200$ OR $10 \times \$350 = \$3,500$.
   The last is a fairly bad estimate, as the exact cost is slightly over $4,000.

2. a. 69,000 b. 408,000 c. 700,000 d. 470,000

3. a. 83,000 b. 660 c. 4,500

4. a. 7 b. 10 c. 9 d. 12

5. a. 277,268 b. 853,758 c. 817,563

Puzzle corner: Twenty-two students brought 9 popsicle sticks and four students brought 7.

Skills Review 29, p. 35

1. a. 9, 3 b. 7, 2 c. 5, 12 d. 8, 9

2. a. $x + 25 = 68$; $x = 43$ b. $92 - x = 74$; $x = 18$

3. $57 \div 6 = 9$ R3. The pigeon got three peanuts.

4. a. 6,920 b. 3,600 c. 1,991

5. Answers will vary. Please check the student's work. See examples below:
   a. $\$750 + \$1,600 = \$2,350$ or $\$800 + \$1,600 = \$2,400$
   b. $4 \times \$9 = \$36$ whereas 5 cups would cost more than $40. You can buy four coffee cups.

6. 3,247; $3247 + 3960 + 823 = 8,030$

Skill Review 30, p. 36

1. $(32 + 45 + 18) \times 10 = 950$. They sold a total of 950 granola bars.

2. Please check the student's answer. $9 \times 200 + 9 \times 80 + 9 \times 4 = 2,556$.

200 80 4
9

3.

Hair Color	Frequency
blond	8
black	5
brown	6
gray	6
red	3

a.

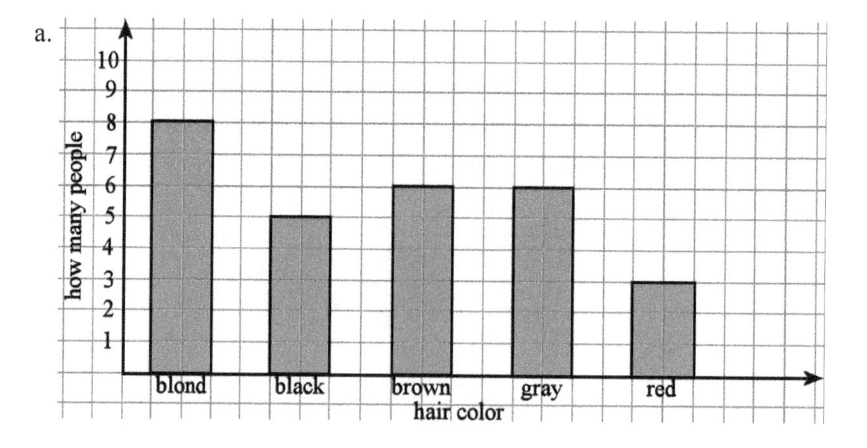

b. red
c. 28 people

4. a. 80 b. 52 c. 330

Skills Review 31, p. 37

1. a. 250,000; 210,000 b. 1,600; 2,400,000 c. 810,000; 132,000

2. a. Discount: \$19.55 b. Before: \$72.33

3. a. $4 \times 37 + 20 = 168$; They collected a total of 168 seashells.
   b. For example: $20 \times \$12 = \$240$; $18 \times \$12 = \$216$; $16 \times \$12 = \$192$; He would have to work 16 hours.

4.

number	468,237	216,951	847,623	566,315
to the nearest 1,000	468,000	217,000	848,000	566,000
to the nearest 10,000	470,000	220,000	850,000	570,000
to the nearest 100,000	500,000	200,000	800,000	600,000

Skills Review 32, p. 38

1. 8,661

2. a. 873 b. 258 c. 201 d. 232

3. a. $x = 48$ b. $x = 97$ c. $x = 454$

4. a. $7 \times 8 + 4 \times 9 + 6 = 98$. She picked 98 flowers.
   b. $60 \div 7 = 8$ R4 or $7 \times 8 = 56$. She will have eight full rows, and four plants in the partial row.

Skills Review 33, p. 39

1. $x = \$166.38$

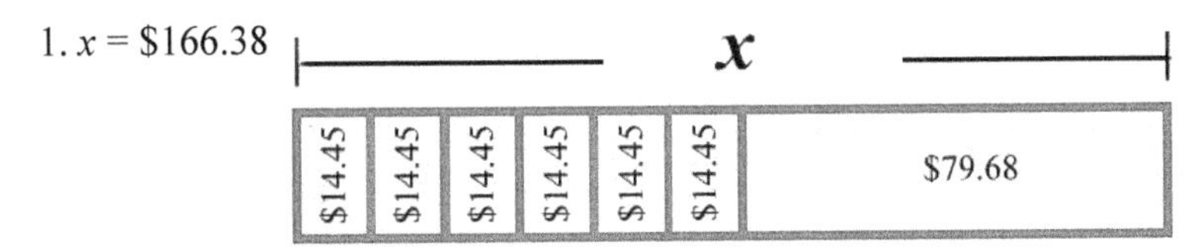

2. a. 15,478 b. 546,423

3.

a. $7 \times 88 \approx 7 \times 90 = 630$	b. $35 \times 243 \approx 40 \times 240 = 9{,}600$	c. $692 \times 9 \approx 690 \times 10 = 6{,}900$

4. a. April, May, October and December
   b. January and September
   c. No. Paris does not have any month where the rainfall would be zero or near zero.

Skills Review 34, p. 40

1.

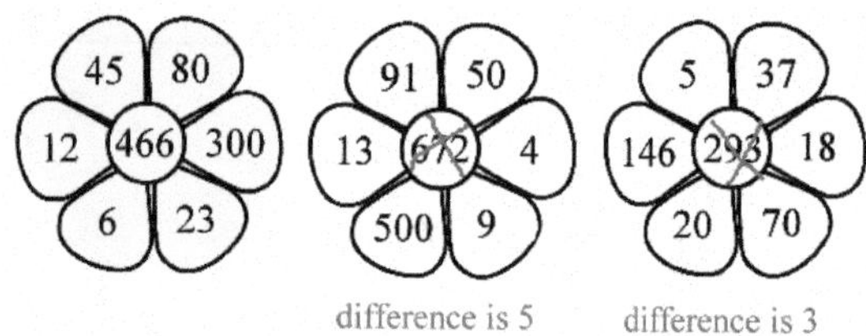

2. a. < b. < c. >

3. a. 100 b. 10 c. 1,000

4.

Words	45	90	135	180	225	270	315	360	405	450
Minutes	1	2	3	4	5	6	7	8	9	10

5. She can buy twelve packages of blueberries.

6. $12.65 + $6.79 + $7.48 = $26.92; $50 – $26.92; His change was $23.08.

7. a. 5 b. 7 c. 8 d. 11 e. 7 f. 9

Skills Review 35, p. 41

1. a. $47.14; $47.14 + 42.86 = $90.00

   b. $108.25; $108.25 + 391.75 = $500

2. They still need $69.

3. a. Starting from the left side: each number on the right is one-half of the number on the left.

   b. Starting from the left side: each number on the right is double the number on the left.

4.

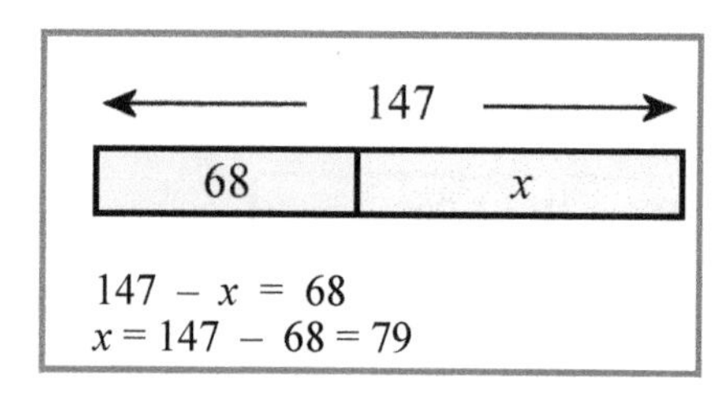

5. a. 12 b. 9 c. 8 d. 0

6. The total area is 2,553.

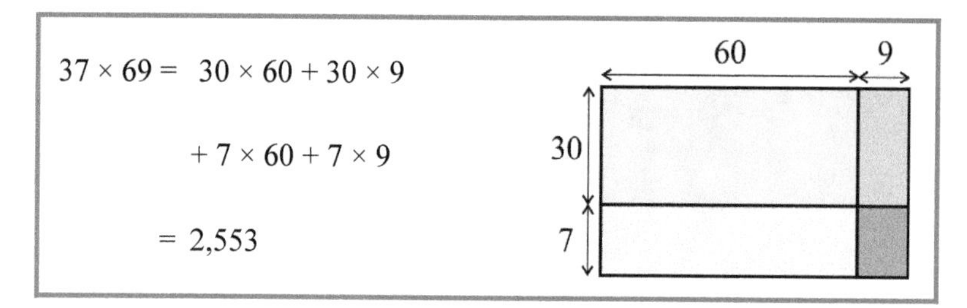

Chapter 4: Time and Measuring

Skills Review 36, p. 42

1. a. 4,740 b. 2,460 c. 20,280 d. 13,140 e. 6,500

2. $\$885 + x = 929$ or $\$929 - \$885 = x$ or $\$929 - x = \$885$; $x = \$44$. The discount is $44.

⟵ original price $929 ⟶

$885	$44

3.

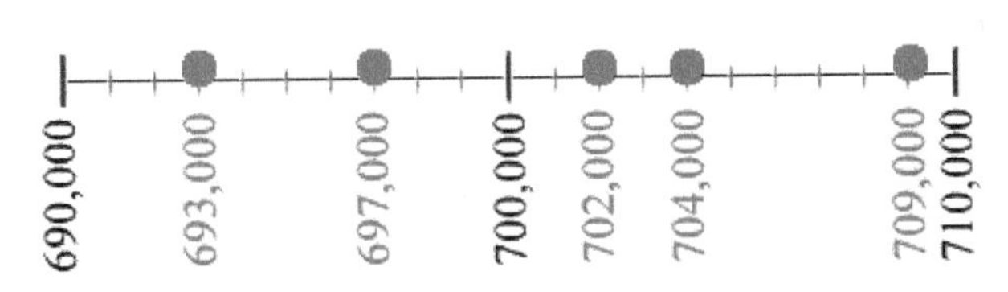

4.

×	0	1	2	3	4	5	6	7	8	9	10	11	12
9	0	9	18	27	36	45	54	63	72	81	90	99	108
4	0	4	8	12	16	20	24	28	32	36	40	44	48
12	0	12	24	36	48	60	72	84	96	108	120	132	144
7	0	7	14	21	28	35	42	49	56	63	70	77	84
11	0	11	22	33	44	55	66	77	88	99	110	121	132
8	0	8	16	24	32	40	48	56	64	72	80	88	96
6	0	6	12	18	24	30	36	2	48	54	60	66	72

Skills Review 37, p. 43

1. a. $90 - 60 = 30$ b. $3,400 - 800 = 2,600$

2.

24 ÷ 3	21 ÷ 7	144 ÷ 12	33 ÷ 11
10 ÷ 2	12 ÷ 2	72 ÷ 3	4 ÷ 4
6 ÷ 2	66 ÷ 11	9 ÷ 3	36 ÷ 6

3. a. 2, 4, 8 b. 2, 4, 7

4.

Number	Rounded number	Rounding error
7,634	8,000	366
5,249	5,000	249
3,768	4,000	232

5. a. Estimate: $10 \times 270 \approx 2,700$ or $8 \times 300 \approx 2,400$. Exact: 2,184
   b. Estimate: $6 \times 600 \approx 3,600$ Exact: 3,564

6. $\$54,235 - (\$6,500 + \$13,000) = \$34,735$. He has $34,735 left in his savings.

Skills Review 38, p. 44

1. a. 2,444 b. 2,226 c. 3,876 d. 2,970

2. a. One triangle = 4. b. One square = 6.

3. a. Greg earns about $8 \times 500 = \$4,000$. Taylor earns about $2 \times 2,800 = \$5,600$.
   b. The difference is about $1,600.

4. a.

Days	Hours
5	120
6	144
7	168

b.

Minutes	Seconds
6	360
7	420
8	480

c.

Years	Months
9	108
10	120
11	132

Skills Review 39, p. 45

1.

a.	b.	c.	d.
3 h 30 m 2 h 30 m + 1 h 15 m 7 h 15 m	2 h 22 m 4 h 33 m + 1 h 18 m 8 h 13 m	6 h 35 m 2 h 19 m + 1 h 11 m 8 h 5 m	1 h 44 m 3 h 13 m + 5 h 27 m 10 h 24 m

2. a.

1 shirt	$6
7 shirts	$42
8 shirts	$48

b.

1 notebook	$0.50
5 notebooks	$2.50
12 notebooks	$30

3.

a. $40 \times 70 = 2,800$
b. $90 \times 600 = 54,000$
c. $800 \times 300 = 240,000$

4.

a. $84 \div 7 = 12$	d. $63 \div 9 = 7$
b. $108 \div 12 = 9$	e. $140 \div 7 = 20$
c. $56 \div 8 = 7$	f. $48 \div 6 = 8$

5. a. 1 h 20 m + 1 h 20 m + 3 h = 5 h 40 m. It will take him 5 hours and 40 minutes for his shopping trip.
   b. $2 \times 60 - 24 = 96$ seconds. Joan's finishing time was 96 seconds.

Skills Review 40, p. 46

1. Please check the student's drawing. See an example on the right:
   $24 \times 56 = 4 \times 6 + 4 \times 50 + 20 \times 6 + 20 \times 50 = 1,344$.

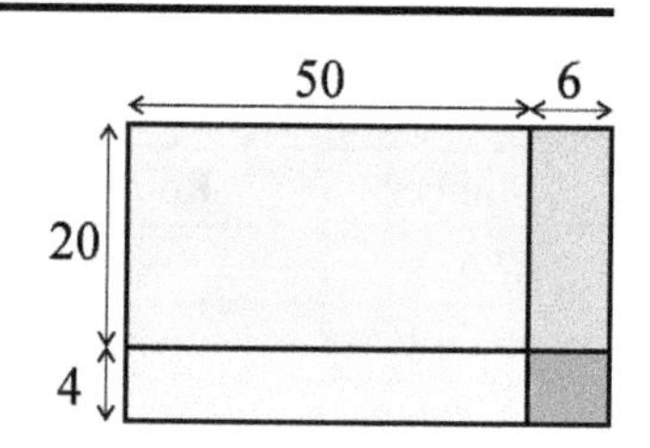

2. 1st square: 3,200 2nd square: 6,800 3rd square: 6,200 4th square: 13,400

3. a. She woke up at 4:20 a.m.
   b. He practices 21 hours in two weeks. 2:30 to 4:15 is 1 h 45 min = 105 minutes.
   $2 \times 6 \times 105 = 1,260$ minutes; $1,260 \div 60 = 21$ hours.

4. Estimates will vary. For example: a. $250 \times \$4 \approx \$1,000$ or $300 \times \$3.50 \approx \$1,050$ b. $600 \times \$1.20 \approx \$720$

Skills Review 41, p. 47

1. a. January and May b. about $6,000

2. a. 786,420 b. 552,675

3. a. 0° C b. 100° C c. 37 d. Answers will vary.

4. a. $8 \times \$27.56 = \$220.48$ b. $4 \times \$83.14 = \$332.56$

Skills Review 42, p. 48

1.
8,600
7,700
6,800
5,900
5,000

2.

n	73	3,453	6,786
nearest 10	70	3,450	6,790
nearest 100	100	3,500	6,800
nearest 1000	0	3,000	7,000

3. They worked a total of $5 \times 53 + 7 \times 45 = 580$ hours.

4. $7{,}028 < 7{,}082 < 7{,}208 < 7{,}280 < 7{,}802 < 7{,}820$

5. a. $9 \times 6 = 54$; $11 \times 6 = 66$ b. $7 \times 4 = 28$; $12 \times 11 = 132$;
c. $8 \times 8 = 64$; $12 \times 4 = 48$; d. $11 \times 11 = 121$; $7 \times 7 = 49$

6. Please check the student's answers. Answers will vary. For example:
a. A very hot summer day. b. A cool fall day.

7. a. 700; 585 b. 453; 86 c. 90; 6,200

Skills Review 43, p. 49

1. a. 7:38 p.m. b. 6:50 a.m. c. 9:45 p.m. d. 1:29 p.m.

2. a. 7 b. 6 c. 7 d. 6 e. 7 f. 7

3. a. Estimation: $7 \times \$9 = \$63$ or $7 \times \$9.40 = \$65.80$ Exact: $66.01
b. Estimation: $9 \times \$80 = \$720$ Exact: $729.45
c. Estimation: $6 \times \$50 = \$300$ or $6 \times \$47 = \$282$ Exact: $280.32

4. She spends a total of $45 \times 14 = 630$ minutes, which is 10 hours and 30 minutes, walking her dog.

5. $\$345 + x = \$520$; $x = \$520 - \$345$. He earned $175.

← total x →

$345	$175

Skills Review 44, p. 50

1. a. 1,950 b. 560 c. 66,060 d. 56,140 e. 27,920

2. a. You could buy four coloring books.
b. Answers will vary. For example: The total cost was about $3 \times \$3.50 + 2 \times \$6 = \$10.50 + \$12 = \$22.50$.

3. Please check the lines the student drew.

4.

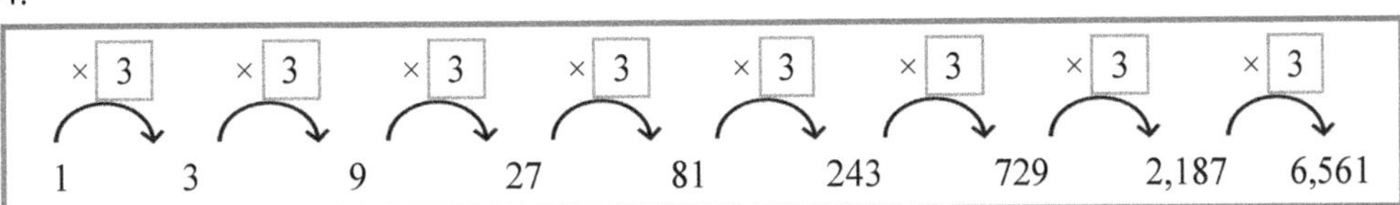

5. a. $7{,}000 < 7{,}200$ b. $180 = 180$ c. $405 > 400$

6. She walks $6 \times 4 \times 7 = 168$ miles in seven weeks.

Skills Review 45, p. 51

1.

number	645,498	462,784
to the nearest 1,000	645,000	463,000
to the nearest 10,000	650,000	460,000
to the nearest 100,000	600,000	500,000

2.

a. 8 ft = 96 in 12 ft = 144 in
b. 72 in = 6 ft 90 in = 7 ft 6 in

3. 4,728 + 7 × 877= 4,728 + 6,139 = 10,867

4. a. There are 6 × 472 = 2,832 people in Sunny City. b. She earns $120 ÷ 8 × 7 = $105 in seven hours.

Skills Review 46, p. 52

1. One circle equals 6. $x + x + x = 12 + x$; $x + x = 12$. One half of 12 is 6 so one x or circle equals 6.

2. a. 163 b. 729

3. She finished at 6:15 p.m.

4. a. 28 oz = 1 lb 12 oz 39 oz = 2 lb 7 oz	b. 48 oz = 3 lb 0 oz 41 oz = 2 lb 9 oz	c. 55 oz = 3 lb 7oz 65 oz = 4 lb 1 oz

5. a. 11 twigs b. 6 5/8 inches c. The three longest twigs are 5 2/8 + 6 2/8 + 6 5/8 = 18 1/8 inches long.

6. a. 9 b. 7 c. 5 d. 6 e. 9 f. 7

7. a. Their total combined weight was: 16 × (3 + 5) + 12 + 14 = 154 ounces. b. 9 lb 10 oz.

Skills Review 47, p. 53

1. Estimations will vary. Check the student's estimation. The following estimations show one possibility.
   a. Estimate: 50 × 60 ≈ 3,000. Exact: 2,976. b. Estimate: 90 × 30 ≈ 2,700. Exact: 3,003
   c. Estimate: 60 × 20 ≈ 1,200. Exact: 1,003 d. Estimate: 20 × 80 ≈ 1,600. Exact: 1,748

2.

9,300 8,500 7,700 6,900 6,100 5,300

3.

a. 9 m = 900 cm 14 m = 1,400 cm
b. 7 m 8 cm = 708 cm 12 m 30 cm = 1,230 cm
c. 900 cm = 9 m 548 cm = 5 m 48 cm

4. a. x = $27.95 + $17.60 + $287 or x − $27.95 − $17.60 = $287; x = $332.55
   b. $81.04 + x = $106 or $106 − x = $81.04; x = $24.96

Skills Review 48, p. 54

1. a. 12 b. 3 c. 12

2. a. 8 oz or 4 oz b. 2T c. 90 lb or 150 lb

3. Drawings may vary; see the example on the right.

   3 × 5 + 3 × 60 + 80 × 5 + 80 × 60 = 5,395

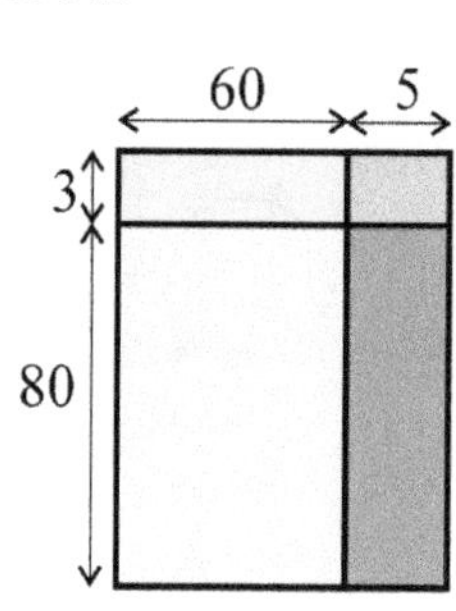

4. a. 70,000 + 9,000 + 600 + 20 + 8
   b. 300,000 + 40,000 + 2,000 + 400 + 10 + 5

Puzzle Corner: a. x = 42 b. x = 197

Chapter 5: Division

Skills Review 49, p. 55

1. a. 151,457 b. 1,515,252 c. 156,782

2. a. 7,000 g; 4,000 g b. 2,300 g; 9,050 g c. 6,200 g; 3,002 g

3. a. 1,000 b. 63,000 c. 160,000

4. a. One diamond weighs 7. b. One circle weighs 6. One triangle weighs 9.

5. The discount is $7.50. ⟵ original price $67.45 ⟶

$59.95	x

Skills Review 50, p. 56

1. a.

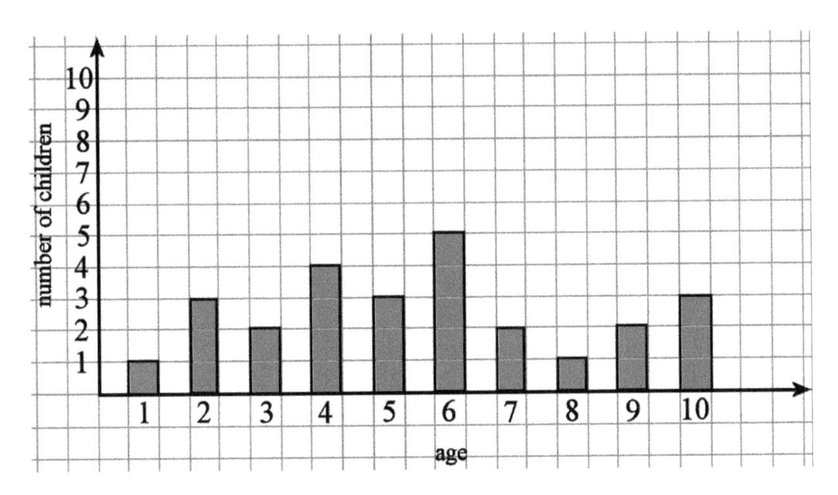

Age	Frequency
1	1
2	3
3	2
4	4
5	3
6	5
7	2
8	1
9	2
10	3

b. 20 c. 6

2. a. 3,840 + 160 = 4,000 b. 7,310 + 690 = 8,000 c. 5,490 + 510 = 6,000

3. a. 7 × 50 = 350 b. 20 × 50 = 1,000 c. 200 × 12 = 2,400

4. a.

Quarts	Cups
7	28
9	36

b.

Quarts	Pints
3	6
8	16

c.

Gallons	Quarts
1/2	2
6	24

Skills Review 51, p. 57

1. a. 22 hours b. 14 hours 57 minutes c. 18 hours 39 minutes d. 15 hours 1 minute

2. a. 8,000 ml; 3,000 ml b. 12 L; 4 L c. 6,095 ml; 10,460 ml

3.

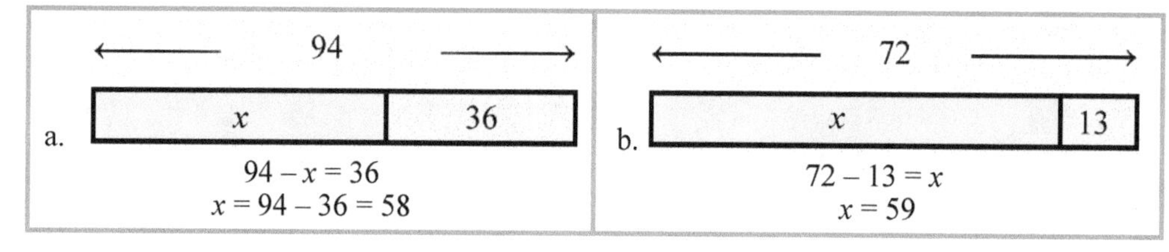

4. a. 80; 30 b. 1,000; 430 c. 600; 21,000

5.

Dollars	$1.60	$3.20	$4.80	$6.40	$8.00	$9.60	$11.20	$12.80	$14.40	$16.00
Yards	1	2	3	4	5	6	7	8	9	10

Skills Review 52, p. 58

1. a. 7 b. 6 c. 32 d. 48

2. a. May b. She had $105 more. c. She needs to save $87.

3. a. $10.92 + $49.08 = $60.00 b. $139.80 + $360.20 = $500

4. a. > b. < c. < d. > e. > f. <

5. 200,000; 3,000

6. a. 8,460 b. 3,400 c. 21,490

Skills Review 53, p. 59

1.

n	38	128	218	308	398	488	578	668	758	848
$n - \underline{29}$	9	99	189	279	369	459	549	639	729	819

2.

$9 \times 147 = 9 \times 100 + 9 \times 40 + 9 \times 7 = 1{,}323$

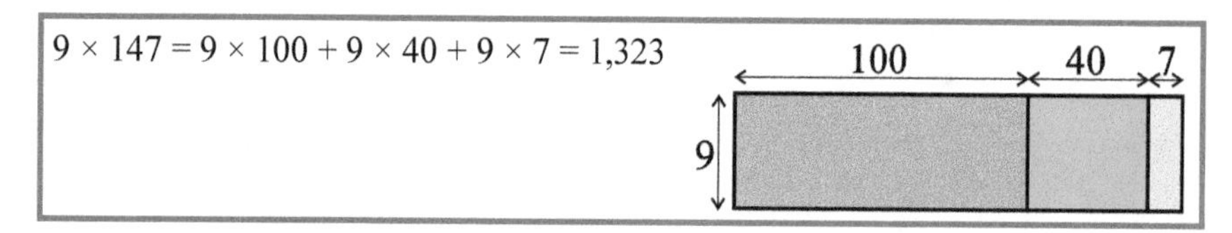

3. a. 6:35 p.m. b. 9:54 p.m. c. 3:44 a.m.

4. Answers will vary. Please check the student's answers.

5. a. 0; 8 b. 5; impossible c. 6; 9 d. impossible; 12

Skills Review 54, p. 60

1. a. 40 b. 9,000 and 1

2. This means there were about 73,000 visitors on Friday, and about 118,000 visitors on Saturday. The two days had approximately 191,000 visitors in all. There were about 45,000 more visitors on Saturday than on Friday.

3. a. Starting from the far left side, and going clockwise:
   1 1/4 in, 2 1/2 in, 2 1/8 in, and 1 5/8 in.
   b. 3 cm 1 mm (or 3 cm 2 mm), 6 cm 4 mm, 5 cm 5 mm, 4 cm 2 mm

4. a. 50; 5 b. 600; 60 c. 9; 900

5. ($1800 – $960) ÷ 120 = 7. She needs to save for seven more weeks.

Skills Review 55, p. 61

1. a. false b. true c. false d. false e. true

2. a. 18, 24 b. 12, 30 c. 14, 42

3. a. 9 × $800 = $7,200
   b. One way to calculate the exact answer is to multiply 9 × $798 = $7,182. Another way is to think of the error of estimation, which is 2 × $9 = $18. From that we find that the total is $7200 – $18 = $7,182.

4. Please check the student's bar model.
   $211 – $182 – $27 = ?
   ? = $2

$211

$182	$27

5.

Half the number	47	175	363
Number	94	350	726
Double the number	188	700	1,452

6.

a. 80 – 36 = 44
b. 200 – 79 = 121
c. 53 – 17 = 36

Skills Review 56, p. 62

1.

×	5	0	7	3	12	4	10	2	9	1	6	11	8
9	45	0	63	27	108	36	90	18	81	9	54	99	72
7	35	0	49	21	84	28	70	14	63	7	42	77	56
8	40	0	56	24	96	32	80	16	72	8	48	88	64
6	30	0	42	18	72	24	60	12	54	6	36	66	48
5	25	0	35	15	60	20	50	10	45	5	30	55	40

2. a. 4:48 p.m. b. 2:13 p.m. c. 10:29 p.m. d. 7:09 p.m.

3. a. 120 b. 100 c. 4

4. a. Brenda is ten inches taller.
   b. He grew 6 inches. When he is 10, he would be 4 ft 8 in. When he is 12, he would be 5 ft 2 in.

5. a. 744,218 b. 258,848 c. 806,251

Skills Review 57, p. 63

1. a. 750,048 b. 84,210

2. a. 3,634 b. 950 c. 5,796 d. 2,280

3. a. 4 R2; 8 R4 b. 7 R3; 11 R2 c. 13 R1; 9 R5

4. a. He rides his bike 325,00 kilometers in thirteen weeks.
   b. Its perimeter is 29 centimeters.

Skills Review 58, p. 64

1. a. 112 oz; 64 oz b. 41 oz; 99 oz c. 133 oz; 63 oz

2. a.

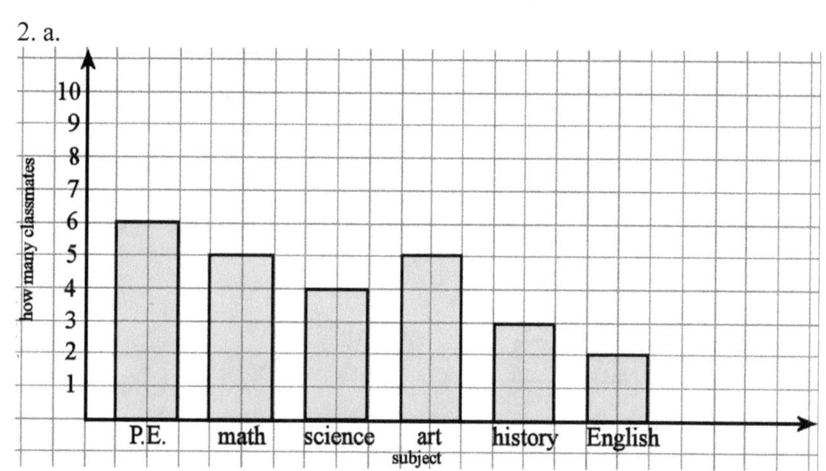

Subject	Frequency
PE	6
Math	5
Science	4
Art	5
History	3
English	2

b. PE c. 25

3. a. $x = 460{,}000$ b. $x = 150{,}000$ c. $x = 45{,}000$

4. a. 0 b. impossible c. 8

5. a. 14,000 b. 300,000 c. 4,800 d. 11 e. 900 f. 7

Skills Review 59, p. 65

1. a. $4 \times 8 - 6 = 5 \times 25 + 1 = 26$
   b. $8 \times 20 = 30 + 130 = 160$
   c. $26 + 24 = 3 \times 16 + 2 = 50$

2. She made a total of 120 ounces of coffee.

3.

Cost	Given	Change	$50 bill	$20 bill	$5 bill	$1 bill
a. $64	$100	$36		1	3	1
b. $28	$50	$22		1		2

4. a. 8 R3 b. 8 R3 c. 11 R2 d. 9 R4

5. a. 420 minutes; 180 minutes
   b. 617 minutes; 767 minutes
   c. 3 hours; 7 hours

6. a. $x - \$16 = \$29$; $x = \$45$ b. $38 + x = 54$; $x = 16$

Skills Review 60, p. 66

1. a. 1/5 part; 60 ÷ 12 = 5 b. 24 cars; 24 ÷ 3 = 8

2. a. 246; 3 × 246 = 738 b. 71; 8 × 71 = 568

3.

n	$44.62	$17.29
to nearest dollar	$45.00	$17.00
to nearest ten dollars	$40.00	$20.00

4. a. 7 kg 60 g b. 90,800 g c. 5 kg 3 g

5.

n	470	520	570	620	670	720	770	820
n – 89	381	431	481	531	581	631	681	731

Skills Review 61, p. 67

1. a. 8 L 315 ml or 8,315 ml b. 9 L 250 ml or 9,250 ml c. 6 L 190 ml or 6,190 ml

2. a. 422; 9 × 422 = 3,798 b. 336; 7 × 336 = 2,532

3. a. 4:05 b. 55 minutes

4. 60 ÷ 10 × 30 = 180; There was a total of 180 sheets of yellow construction paper.

5. a. 40 b. 570 c. 671

Skills Review 62, p. 68

1. a. The week of January 1-7 was snowier.
   b. The average daily snowfall for the first week of January was 5 inches.

2. a. 770,000 b. 329,000 c. 87,400

3. 87 ÷ 6 = 14 R3. She had 14 full packages of cookies.

4. 99 ÷ 8 = 12 R3. They need 13 tables. There will only be 3 people at the thirteenth table.

5. Please check the student's drawing. See an example on the right.

   46 × 29 = 6 × 9 + 6 × 20 + 40 × 9 + 40 × 20 = 1,334.

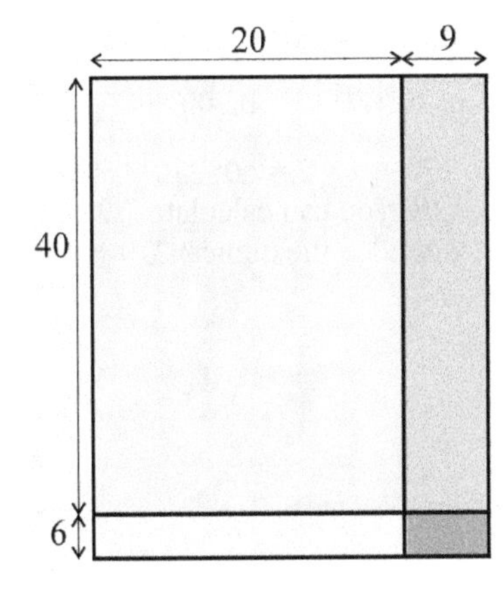

Skills Review 63, p. 69

1. a. 2,765 b. 3,280 c. 77,040 d. 3,564 e. 19,620

2. a. One pint of ice cream costs $1.23
   b. One shirt cost $8.45

3. a. 77 b. 20 c. 820

4. a. 640,000 b. 260,000 c. 640,000 d. 750,000

5. Please check the lines the student drew.

Skills Review 64, p. 70

1. (70 + 8) × 5 = 390 eggs

2. a. Starting with the side that is nearly upright, and going clockwise:
   7/8 in, 1 7/8 in, 2 3/4 in, 1 3/8 in
   b. 2 cm 3 mm, 4 cm 9 mm, 7 cm, 3 cm 6 mm

3. a. 500 ÷ 50 = 10
   b. 750 ÷ 30 = 25
   c. 400 ÷ 80 = 5

4. a. 7; 7 b. 8; 12 c. 7; 7 d. 7; 9

5. 956 are children.

6. a. 21 ft; 54 ft b. 29 ft; 37 ft c. 13 yd; 16 yd

Skills Review 65, p. 71

1. a. 60 × $120 ≈ $7,200 b. 700 × $0.70 ≈ $490

2. a. 70,000 + 4,000 + 200 + 90 + 5
   b. 300,000 + 6,000 + 100 + 80 + 7

3. a. −2° C b. −13° C c. −7° C

4. a. 12 km 400 m b. 22 km 600 m

5. a. 821,711 b. 486,847 c. 1,641,632

6. 270 ÷ 9 × 2 = 60; 270 − 60 = 210.
   OR, you can calculate 7/9 of 270, as that is what part of the employees who attended the dinner: 270 ÷ 9 × 7 = 210. So, 210 employees attended the dinner.

Chapter 6: Geometry

Skills Review 66, p. 72

1. a. $36 \times \underline{100} = 3600$
   b. $71 \times \underline{10} = 710$
   c. $\underline{1,000} \times 53 = 53,000$

2.

12,600	18,900	7,400	11,100	= 50,000
4,500	9,700	21,700	14,100	= 50,000
22,800	6,300	8,700	12,200	= 50,000
10,100	15,100	12,200		
= 50,000	= 50,000	= 50,000		

3. a. 4 lb b. 20 lb c. 150 lb or 240 lb

4. Estimates will vary. Here is one possibility:
   $3 \times \$7 + \$13 + \$14 = \$48$.
   Exact: $3 \times \$6.75 + \$12.89 + \$13.64 = \$46.78$

5. a. 729,140 b. 68,237

Mystery Number: 44, 88, 132

Skills Review 67, p. 73

1. One circle weighs 5. One square weighs 7.

2.

Days	Hours
5	120
3	72
6	144
4	96

Years	Months
7	84
2	24
5	60
3	36

3. a. prime b. composite; $68 = 4 \times 17$ c. composite; $57 = 3 \times 19$

4. They bought 5 gallons of juice. You can first calculate the number of pints, and divide that by 8 to get the amount in gallons: $(2 \times 8 + 10 + 2 \times 7) \div 8 = 5$ gallons. Or, you can first calculate the number of quarts and divide that by 4: $(2 \times 4 + 5 + 7) \div 4 = 5$ gallons

5. a. 188; $188 \times 8 = 1,504$ b. 394; $394 \times 8 = 2,364$

Skills Review 68, p. 74

1. $10,036 < 10,630 < 16,303 < 31,063 < 301,060 < 310,063$

2. $7 \times s = 84$; $s = 12$

3. a. Factors: 1, 2, 3, 4, 6, 8, 9, 12, 18, 24, 36, 72 b. Factors: 1, 2, 3, 6, 9, 18, 27, 54

4. Please check the student's answers. For example: a. A hot summer day. b. A winter day.

5.

$294 + x = 463$
$x = 463 - 294 = 169$

← total 463 →

294	169

6. $\$13.40 \div 4 \times 3 = \$10.05$. Three cans would cost \$10.05.

7. Please check the student's answers. Answers will vary. Examples:
   a. $1 \div 1 = 1$ OR $898 \div 898 = 1$
   b. $0 \div 6 = 0$ OR $0 \div 354 = 0$

Skills Review 69, p. 75

1. a. 6 cm b. 24 cm

2. a. 2:10 p.m. b. 9:18 p.m. c. 10:35 a.m.

3. The total cost was $15,000.

4. There were ,729 more cruise passengers to Key West in March of 2017 than in March of 2016.

5. a. 7,000 g b. 9 kg c. 4,300 g

6. a. $(53 + 71 + 42 + 68 + 56) \div 5$ She received an average of 58 emails.
   b. She received $90 \div 5 \div 3 = 174$ emails.

Skills Review 70, p. 76

1. $x - 4 \times \$7 = \$33$ OR $? = \$7 + \$7 + \$7 + \$7 + \$33 = \$61$. Solution: $x = 4 \times \$7 + \$33 = \$61$

2.

$49 \times 85 = 9 \times 5 + 9 \times 80$
$+ 40 \times 5 + 40 \times 80$
$= 4,165$

80, 5, 40, 9

3.

Child	Marsha	Kyle	Bonnie
Number of cookies baked	76	87	91
How many in each package	8	12	6
Number of full packages	9	7	15
How many cookies left	4	3	1

4. The average price was $18.

5. a. 8,000 ml b. 3 L 430 ml c. 9 L

6. a. segment b. line c. ray

Skills Review 71, p. 77

1. a. 7,020 b. 4,088 c. 48,400 d. 63,910 e. 8,019

2. a. 28 b. 410

3. Please check the student's answers.

4.

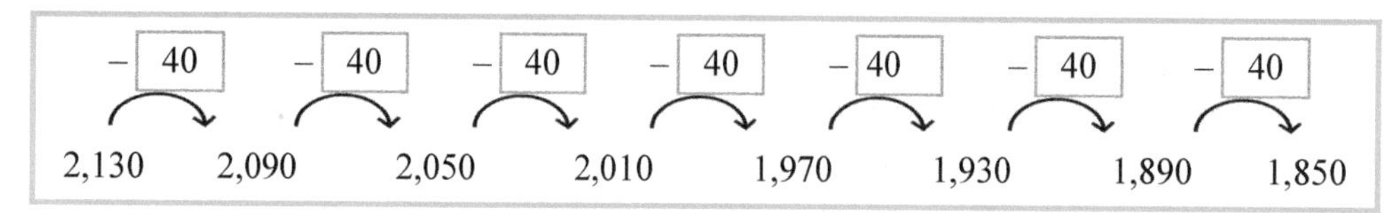

5. Half of the apples is 93 apples. We divide: $93 \div 4 = 23$ R1. So, three of the neighbors get 23 apples and one gets 24.

6. There are 267,418 males and 391,710 females living in Cinnamon City.
   This means there are about 267,000 males and about 392,000 females.
   There are approximately 659,000 people in all.
   There are about 125,000 more females than males in Cinnamon City.

Skills Review 72, p. 78

1.

			91,300			
		36,600		54,700		
	19,800		16,800		37,900	
8,500		11,300		5,500		32,400

2.

a. $\frac{3}{4}$ of 48 is 36.

b. $\frac{2}{6}$ of 54 is 18.

c. $\frac{7}{10}$ of 700 is 490.

d. $\frac{5}{8}$ of 160 is 100.

3. a. Between the ages of nine and ten. b. She weighed 20 pounds more. c. She weighed 49 more pounds.

4. The angle is 158°.

5. a. 60 mm b. 39 mm c. 6 cm 8 mm

Skills Review 73, p. 79

1. Check the student's angles.

a. 43°

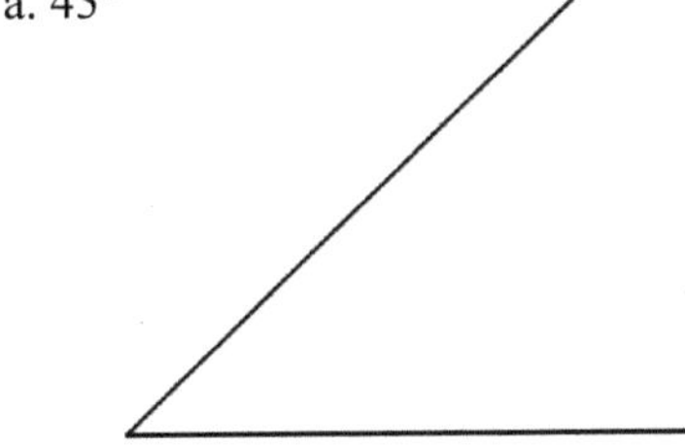

b. 128°

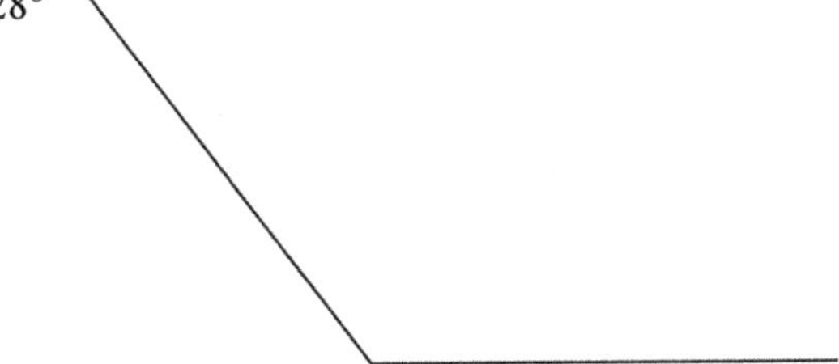

2. About $530 + $140 = $670.

3. They brought a total of 39 × 24 = 936 clothespins, and have 64 fewer than 1,000.

4. a. y = 6,300 b. s = 540 c. w = 6

5. a. b. 11 carrots

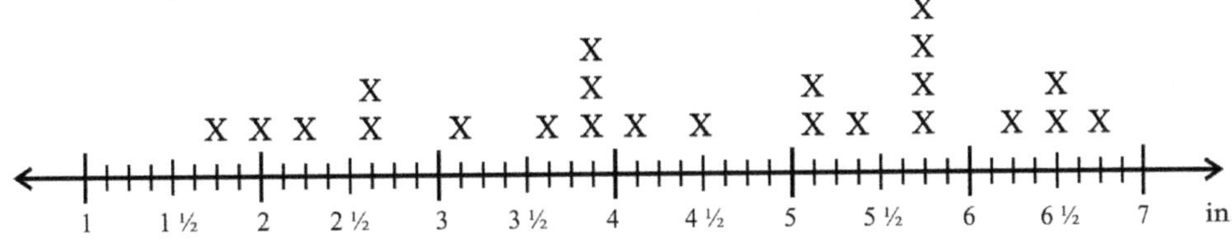

Skills Review 74, p. 80

1. She walked about 2 × 5 × 4,000 ft = 40,000 ft ≈ 8 miles.

2.

128,000	240,000	164,000	320,000
61,000	**32,000**	80,000	82,000
8,000	20,000	15,000	25,000
5,000	**2,000**	7,000	4,000
900	1,250	**500**	1,500

and

128,000	240,000	164,000	**320,000**
61,000	32,000	**80,000**	82,000
8,000	**20,000**	15,000	25,000
5,000	2,000	7,000	4,000
900	**1,250**	500	1,500

3.

×	**3**	**6**	**9**	**12**
0	0	0	0	0
2	6	12	18	24
4	12	24	36	48
6	18	36	54	72
8	24	48	72	96

4. (2 × \$27.60 + \$14.50) ÷ 2 = \$34.85. Each one paid \$34.85.

5. a. ∠A = 115°; ∠B = 85°; ∠C = 130°; ∠D = 30° b. 79°; 154°; 127°

Skills Review 75, p. 81

1. a. 7 × 88 ≈ 7 × 90 = 630 b. 12 × 63 ≈ 12 × 60 = 720 c. 218 × 17 ≈ 200 × 20 = 4,000

2. a. 112 oz; 64 oz b. 108 oz; 147 oz c. 63 oz; 137 oz

3. She can type 3 × (42 − 30) = 36 more words in three minutes than the average person.

4. $l \parallel \overline{AC}$ and $\overline{AB} \perp \overline{BC}$

5. a.

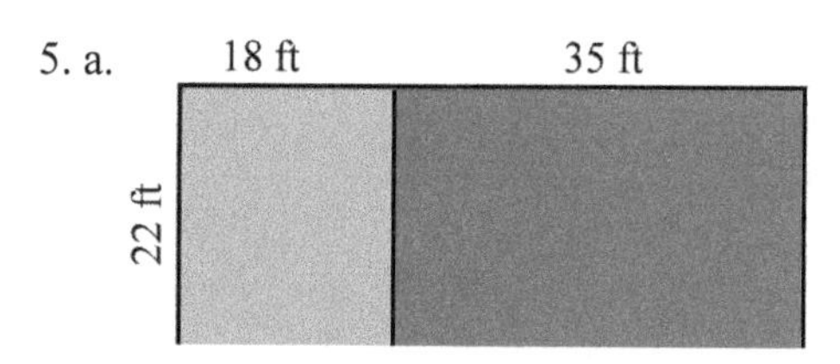

b. Area: 1,166 square foot.

Puzzle corner: a. 20 b. 8

Skills Review 76, p. 82

1. a.9:40 b. 18:00 c. 22:20 d. 4:02

2. a. 480 g ÷ 8 × 3 = 120 g b. One slice is 480 g ÷ 8 = 60 g, so 420 g is seven slices.

3. a - c. Please check the student's answers. The true angle measures are: a. 40° b. 75° c. 135°

4. Sides 6 and 5 units; area 30 square units, perimeter 22 units.

5. a. −31 °C b. −18 °C

6. a. 9 R5 b. 8 R6 c. 12 R3 d. 9 R3

Chapter 7: Fractions

Skills Review 77, p. 83

1. a. 728 ÷ 9 = 80 R8 (an uneven division),
   so 728 is not divisible by 9.
   b. 426 ÷ 6 = 71 (an even division),
   so 426 is divisible by 6.

2.

Minutes	Seconds
2	120
7	420
4	240
9	540
3	180
8	480
5	300
10	600

3. a. trapezoid b. rhombus c. trapezoid d. parallelogram

4. Their total weight was 3 kilograms 920 grams.

5. a. 944,010 b. 236,689 c. 453,184

Skills Review 78, p. 84

1.

163	95	157	117	73
74	151	67	29	169
17	89	191	13	139
173	132	109	63	101
27	128	103	81	121

2. a. 4 C b. 14 pt c. 20 qt d. 12 C

3. a. acute b. obtuse c. right

4. a. 13 hours 4 minutes total:
   5 h 22 min + 7 h 42 min = 13 h 4 min.
   b. 17 hours 14 minutes total:
   9 h + 8 h 14 min = 17 h 14 min.

5. a. 250,000 b. 3,600

6.

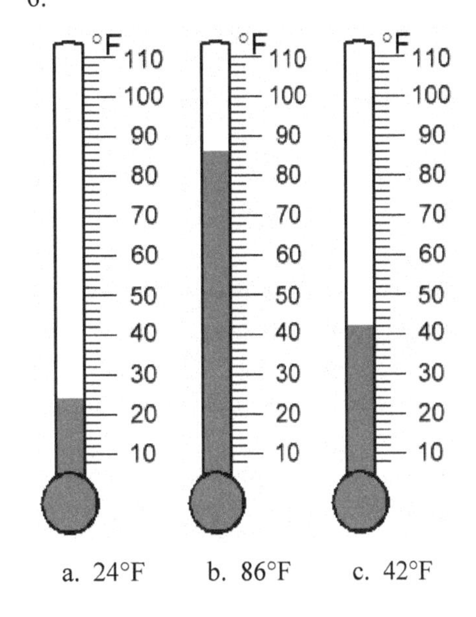

a. 24°F b. 86°F c. 42°F

Skills Review 79, p. 85

1. Please check the student's angles.

a.

b.

Skills Review 79, p. 85, continued

2. a. < b. > c. >

3.

4. She made 4,980 ml or 4 L 980 ml of fruit punch.

5. 1/4, 2 1/4, 3 1/4, 4 1/2, 5 3/4

6. One rectangle equals 8. One pentagon equals 12.

7. a.

b.

c. Not symmetric

d.

Skills Review 80, p. 86

1. To get 48 granola bars, we need four packages. Estimate: 4 × $4.00 = $16.00 Exact: 4 × $3.83 = $15.32.

2.

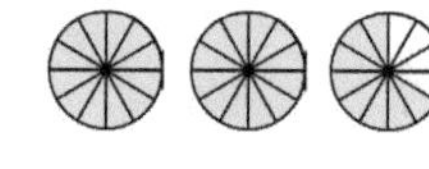

a. $\frac{32}{12} = 2\frac{8}{12} = 2\frac{2}{3}$

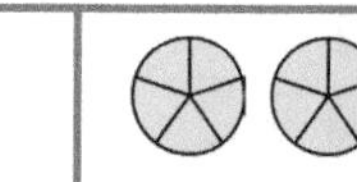

b. $\frac{18}{5} = 3\frac{3}{5}$

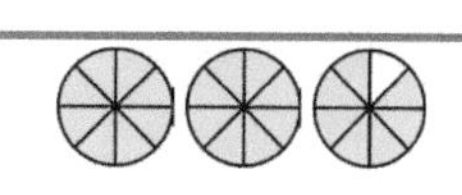

c. $\frac{23}{8} = 2\frac{7}{8}$

3. Please check the student's answer.

4.

Tons	5	7	9	11	16	18	20
Pounds	10,000	14,000	18,000	22,000	32,000	36,000	40,000

5. a. 16 km 200 m b. 12 km 100 m

6. $798 + $1,736 = $2,534

total $2534

$1,736	$798

Skills Review 81, p. 87

1. Please check the student's answer. Answers will vary.

2.

Dollars	\$1.58	\$3.16	\$4.74	\$6.32	\$7.90	\$9.48	\$11.06	\$12.64	\$14.22	\$15.80
Bottles	1	2	3	4	5	6	7	8	9	10

3.

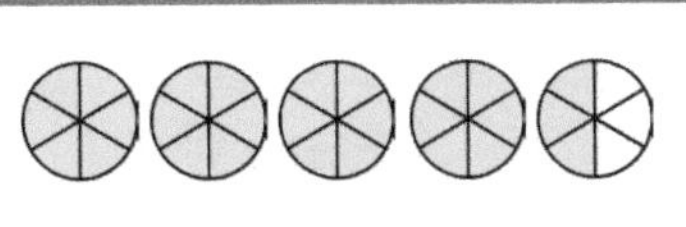	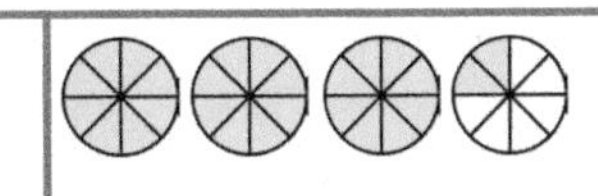
a. $2\frac{5}{6} + 1\frac{4}{6} = 4\frac{3}{6} = 4\frac{1}{2}$	b. $1\frac{3}{8} + 1\frac{7}{8} = 3\frac{2}{8} = 3\frac{1}{4}$

4. The other side is 14 cm long.
   7 + ? + 7 + ? = 42 OR 7 + ? = 21.
   Solution: ? = 14.

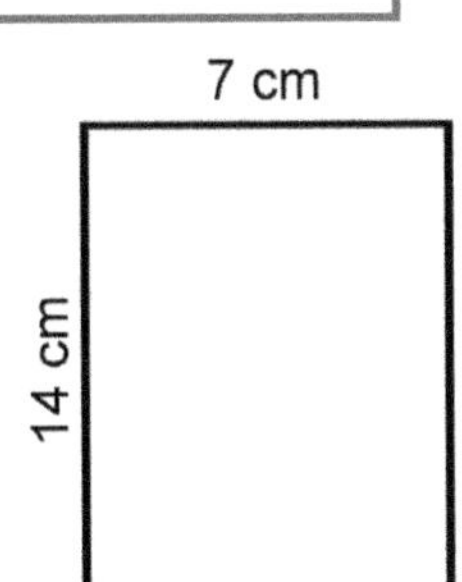

Skills Review 82, p. 88

1. a. Discount \$17.61 b. Original price \$22.13

2. a. angle CAT b. ray ST

3. a. 2 inches is longer b. 7 inches is longer c. 12 cm is longer

4.

a.	b.	c.	d.
	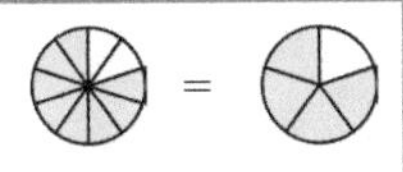	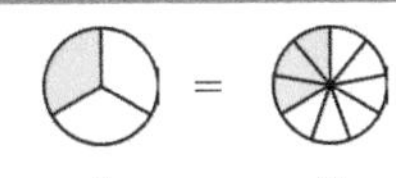	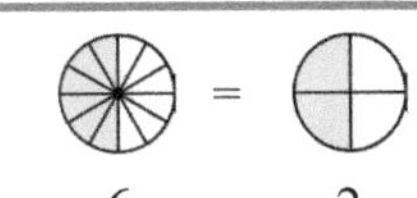
a. $\frac{4}{6} = \frac{8}{12}$	b. $\frac{8}{10} = \frac{4}{5}$	c. $\frac{1}{3} = \frac{3}{9}$	d. $\frac{6}{12} = \frac{2}{4}$

5. a. 80 b. 8 c. 800

6. a. 0 b. 4,400

7.

Round the number	276,302
. . . to the nearest 1,000	276,000
. . . to the nearest 10,000	280,000
. . . to the nearest 100,000	300,000

Skills Review 83, p. 89

1. a. 700 – 4 = 696
   700 – 40 = 660
   700 – 400 = 300
   700 – 44 = 656

2. a. not possible b. 8 c. 0 d. 1 e. 0 f. not possible

3. a. 2,350 b. 2,432 c. 54,720 d. 21,160 e. 3,534

4. a. 24 ft b. 14 ft c. 9 yd

5. (4 + 0 + 3 + 5 + 5 + 7) ÷ 6 = 4. She read an average of four books per week.

6. They drove a total of 349 + 4 × 349 = 1,745 miles.

7.

a.	b.	c.	d.
$2\frac{3}{5} - \frac{4}{5}$	$1\frac{5}{8} - \frac{7}{8}$	$1\frac{4}{10} - \frac{6}{10}$	$2\frac{3}{12} - \frac{9}{12}$
↓ ↓	↓ ↓	↓ ↓	↓ ↓
$\frac{13}{5} - \frac{4}{5} = \frac{9}{5} = 1\frac{4}{5}$	$\frac{13}{8} - \frac{7}{8} = \frac{6}{8}$	$\frac{14}{10} - \frac{6}{10} = \frac{8}{10}$	$\frac{27}{12} - \frac{9}{12} = \frac{18}{12} = 1\frac{6}{12}$

Skills Review 84, p. 90

1. a. June, July, August and September b. 40 degrees c. January, February, March and December

2. a. 75 ÷ 8 = 9 R3; She had nine full rows of chairs with three chairs left over.
   b. 66 ÷ 7 = 9 R3; They needed ten minivans.

3. a. > b. > c. < d. <

4. Check the student's quadrilateral and measurements. One should get very close to 360° if measuring accurately.

Chapter 8: Decimals

Skills Review 85, p. 91

1. a. 15:25 b. 21:40

2.

n	\$74.29	\$8.62
rounded to nearest ten cents	\$74.30	\$8.60
rounded to nearest dollar	\$74	\$9.00

3. Answers will vary. Please check the student's answers.

4. Please check the student's drawings.

a.

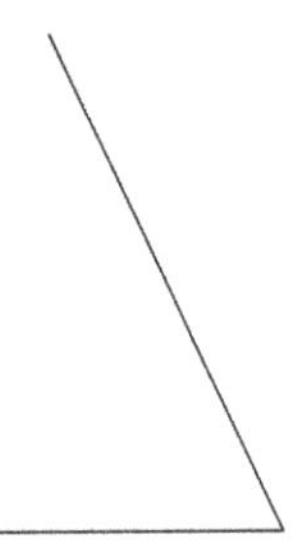

b.

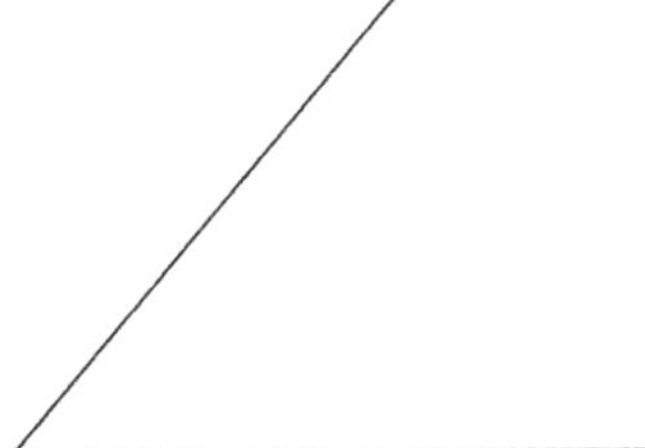

5. a. Yes, because the division is even: $504 \div 7 = 72$. b. No, because the division is uneven: $2{,}748 \div 9 = 305$ R3.

6. a. $\frac{4}{10} = 0.4$	b. $6\frac{7}{10} = 6.7$	c. $8\frac{3}{10} = 8.3$	d. $0.8 = \frac{8}{10}$	e. $32.6 = 32\frac{6}{10}$

7. She typed an average of 24 words per minute.

8. a. $7 \times \frac{2}{5} = \frac{14}{5} = 2\frac{4}{5}$	b. $4 \times \frac{5}{8} = \frac{20}{8} = 2\frac{4}{8}$	c. $6 \times \frac{3}{4} = \frac{18}{4} = 4\frac{2}{4}$	d. $9 \times \frac{4}{7} = \frac{36}{7} = 5\frac{1}{7}$

Skills Review 86, p. 92

1.

Number	Rounded number	Rounding error
6,748	7,000	252
9,591	10,000	409

Number	Rounded number	Rounding error
12,327	12,000	327
4,635	5,000	365

2. $\$3 \times 14 = \$42$ OR $\$3.50 \times 14 = \$49$

3. The difference in length is 3 ft 8 inches.

4. $79 \div 4 = 19$ R3. This means he put 19 cows in one pasture and 20 cows in each of three pastures

5. $56° + x = 90°$; $x = 34°$.

6. a. 3.3 b. 0.7 c. 8.3 d. 2.8

7. a. $\frac{3}{12} + 1\frac{9}{12} = 2$
b. $1\frac{5}{9} + 1\frac{6}{9} = 3\frac{2}{9}$
c. $5\frac{4}{7} + 3\frac{2}{7} = 8\frac{6}{7}$

Skills Review 87, p. 93

1. a. 110°, 30°, 40°   b. 20°, 70°, 90°

2. \$72 ÷ 8 × 7 = \$63. It would cost \$63 for seven pillows.

3. a. >   b. =   c. <   d. =   e. <   f. <

4.

$\frac{2}{3}$	**$\frac{6}{12}$**	$\frac{1}{4}$	$\frac{4}{10}$
$\frac{3}{8}$	$\frac{1}{3}$	**$\frac{1}{2}$**	$\frac{8}{12}$
$\frac{2}{7}$	**$\frac{4}{8}$**	$\frac{7}{9}$	$\frac{3}{5}$
$\frac{2}{4}$	$\frac{6}{10}$	$\frac{4}{12}$	$\frac{1}{3}$
$\frac{3}{7}$	**$\frac{3}{6}$**	$\frac{5}{8}$	$\frac{7}{10}$
$\frac{4}{10}$	$\frac{2}{3}$	**$\frac{5}{10}$**	$\frac{6}{11}$

5.

Now		After
a. −8° C	temperature rises 5° C	−3° C
b. 2° C	temperature falls 8° C	−6° C
c. −7° C	temperature rises 6° C	−1° C

Skills Review 88, p. 94

1. a. 26,000   b. 340,000   c. 820,000   d. 720,000

2. a. 81 R7; 81 × 9 + 7 = 736   b. 76 R5; 76 × 6 + 5 = 461

3. AE and BC are parallel; AB and DE are parallel; AB and ED are perpendicular; DE and EA are perpendicular

4.

a. $5\frac{3}{8} - \frac{7}{8} = 4\frac{4}{8} = 4\frac{1}{2}$	b. $3\frac{2}{5} - \frac{3}{5} = 2\frac{4}{5}$	c. $7\frac{1}{10} - 4\frac{6}{10} = 2\frac{5}{10} = 2\frac{1}{2}$

5. She started at 2:50 p.m.

6.

a.	b.	c.
1.3 − 0.6 + 2.7 = 3.4	8.2 + 1.5 − 7.9 + 2.4 = 4.2	6.4 + 3.9 − 5.4 = 4.9